All About Me

Things I See

By Meg Gaertner

level 1 little blue readers

www.littlebluehousebooks.com

Little Blue House is distributed by North Star Editions:
sales@northstareditions.com | 888-417-0195

Produced for Little Blue House by Red Line Editorial.

Photographs ©: Dragon Images/Shutterstock Images, cover; Vesnaandjic/iStockphoto, 4; aldomurillo/iStockphoto, 7; Steex/iStockphoto, 8–9; FlairImages/iStockphoto, 11; kali9/iStockphoto, 12–13; gollykim/iStockphoto, 15; Tramino/iStockphoto, 16 (top left); xxmmxx/iStockphoto, 16 (top right); skynesher/iStockphoto, 16 (bottom left); Veronika 7833/Shutterstock Images, 16 (bottom right)

Library of Congress Control Number: 2019908620

ISBN
978-1-64619-005-8 (hardcover)
978-1-64619-044-7 (paperback)
978-1-64619-083-6 (ebook pdf)
978-1-64619-122-2 (hosted ebook)

Printed in the United States of America
Mankato, MN
012020

About the Author

Meg Gaertner enjoys reading, writing, dancing, and being outside. She lives in Minnesota.

Table of Contents

dog

Things I See

I see a dog.

I see a cat.

cat

I see a book.

book

I see a car.

car

I see a swing.

swing

I see a cloud.

cloud

Glossary

car

cloud

cat

dog

Index